A Unified Model
of the Human Psyche

A Unified Model of the Human Psyche
by John-Paul Miller

PUBLISHED BY

LuLu Enterprises, Inc.
860 Aviation Parkway, Suite 300
Morrisville, NC 27560
(919) 459-9858

ISBN: 978-1-4357-0067-3
Printed and Bound in the United States of America.

CREDITS

Author:	John-Paul Miller
Illustrations:	John-Paul Miller

A Unified Model of the Human Psyche

(Concise Presentation)

by
John-Paul Miller

Contents

Introduction

A Unified Model of the Human Psyche

This is a concise presentation of a unified model of the human psyche as it pertains to the process of personal growth.

The model integrates individual identity, acceptance and denial, neuroses, pleasure and pain, self-esteem, perception and presentation, love, success, and failure into a unified framework showing the relationships between these elements and providing insight into how to manage them.

The model illustrates and explains the mechanisms by which the inner workings of the human psyche operate. As such it provides an underlying framework for understanding and applying the work of other authors, therapists, counselors, clergy, psychologists, and theorists. Rather than being considered as a replacement for other works, this model should be used as a tool for understanding and applying them.

Having been developed from a purely practical perspective, seeking always to describe "how this thing really works," the model is a usable, practical model, suited to universal application.

Beyond the fundamental need to find and accept truth and understanding within oneself, the model does not assume any particular point of view, philosophy, or religion.

The model can be used as a tool for understanding, as a tool for exploring and personal growth, and also a tool for designing treatments and therapies. It can be applied inwardly to understand and improve oneself, and also outwardly to understand others and to improve interactions with them.

The model is a logical or functional model, not a physical or implementation model. Therefore, it deals with psychology and sociology, rather than anatomy and physiology. Chemical causes are not addressed, and proper physiological function is assumed as a baseline for application. Physical addiction, chemically caused depression, and other illnesses caused by chemical or physical problems are outside the scope of the model. Emotional addiction, emotional depression, and emotional problems, however, are well within the boundaries of the model.

One basic premise is that treatment of a properly functioning physiology with drugs should be avoided. Drugs should be used only in instances where a clear chemical or physiological deficiency is identifiable, seeking instead to treat the psyche through psychology, rather than altering the physical brain function.

As a unified model, it integrates the various schools of psychological and philosophical thought into a common whole for application. It does not constrain the individual in terms of which philosophy is right or wrong, which religion is good or bad, what values are ideal. These elements are unique to the individual, and must be discovered for each person individually.

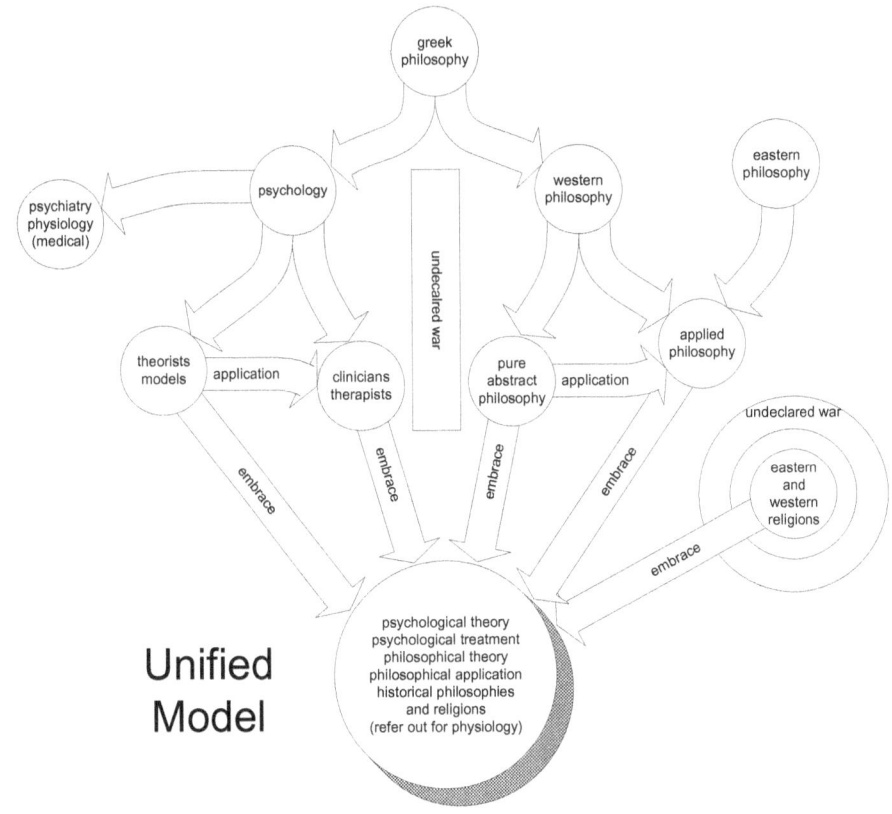

Historically, while psychology and philosophy have common origins, they have developed along widely divergent paths. Psychology has been focused almost exclusively on fixing broken psyches, with pure theorists explaining why they are broken and clinicians building on these explanations and developing treatment regimens to be applied formulaically "by the numbers". Philosophy has been focused on finding the intellectual truths of the world (usually at direct odds with the organized religions), with pure theorists explaining how to find the truth, and practical philosophers building complete sets of what is true about the world. Meanwhile the most of the world's religions were persecuting anyone with a philosophy contrary to their own. This model embraces both psychology and philosophy as two halves of a greater whole, rather than as two separate competing wholes.

The model also fully embraces religion as a valid source of personal philosophy and identity.

This model explains how the psyche works in mechanical terms. Yet, in the absence of any skill at applying philosophical methods, knowing how it works is of limited benefit. The biggest obstacle is that we tend to try to help people by making them exactly like ourselves. To extend Descartes, "We think, therefore we believe our thoughts are the ultimate authority."

Yet, each of us ultimately is different, and has to discover our own particular self-truths. Therefore, philosophical methods have to be utilized by each individual in order to choose the right paths for growth. And each individual, having truths discovered and incorporated into their personal philosophy, ultimately also has to integrate into them into the actual workings of their psyche. Therefore, the application of psychology is also important.

This model integrates the two halves, bringing philosophy and psychology together into a single model which not only explains how a person works, but also provides tools for discovering that which is true for each given individual.

The elements of the model are presented as simple statements of fact, with the minimum explanation of meaning, and with relationships to other elements of the model. These are intended as a quick reference guide to the model and the theories on which it is based. Notably absent from this concise presentation are arguments to substantiate the model and examples to illustrate it. The objective is to avoid getting distracted with arguments and to focus instead on presenting the complete model in its simplest, clearest form. Further explanation and discussion is intentionally left to be provided separately in a much longer, more descriptive presentation.

Many of the statements of fact will seem absurd at first glance, begging for an immediate argument. Keep in mind that no single statement of fact is intended to stand alone, however, reestablishing the context of the model for each statement would make for tedious reading. So while all arguments and challenges are encouraged, try to focus on the context of the model as a whole.

As you work through the various elements of the model, you are encouraged to consider how the model applies to other theories, teachings, examples, and personal experiences. This model is not expected to be the final version as future discoveries and revelations can contribute to improving the model. Nor does it intend to present a set of ultimate truths as those are unique to each individual.

It should be stated that the model should not be improved by tacking on new rooms (disorders and treatments), as is often done with the DSM. Rather, the model should be reengineered in light of new discoveries, maintaining elegant simplicity and universality as the ultimate goal. If a new discovery can be decomposed and shown to be described by the model, then it should become an addendum work of application, serving to demonstrate its application, while also validating the model itself.

Many of the examples are given as examples of physical senses and actions. By demonstrating the better understood and more easily demonstrated physical patterns, emotional patterns can be illustrated by analogy.

Application of theory is an art, one that each must master in a unique way in order to derive the full benefits promised by the theory itself.

Fundamental Model - Components of Identity

An identity comprises a set of identity elements

Our identity is not a single thing, rather is it a collection of a diverse set of facts about us. It is the complete set of facts that describes every detail about us, everything we experience, everything we value, and everything we know about the world around us.

One key concept that is critical for the universal application of this theory is the broad definition of identity as including every single thing that defines a person, with each self truth able to be expressed as a different and distinct identity element which can be examined in the context of the model.

Thousands of books have been written addressing various personal crisis situations, and the DSM has documented hundreds of unique variations of personal crises. But, once the scope of identity is broadened and defined as the collection of facts that describes every single thing about a person, then all these different crises can be understood for what they really are: Various forms of identity crises.

Absent physical causes, every emotional issue is actually a varying degree of identity crisis affecting a different subset of a person's identity elements.

Each identity element is a self-truth

Each identity element is a single fact that describes a person.

A person's entire identity is the complete set of facts that describes them.

Identity can never be fully known

A person continues to discover and explore his or her sense of identity. We never fully understand ourselves. We live a continuous process of discovery and evolution of our identity.

Identity continues to evolve

As long as a person lives, he or she continues to learn through new experiences and new discoveries, and these experiences and discoveries lead to new insights about his or her identity.

Each crisis leads to an evolution of personal identity, as the process of perception, acceptance, and mastery continues.

The greater the extent to which a person is able to reexamine, evolve, and accept a changing sense of identity, the more that person will grow and find happiness, personal success, and emotional peace.

One goal is to learn to evolve identity without requiring the trigger of an actual crisis.

Types of identity elements

There are many different types of individual identity elements that make up the whole of a person.

Values – The moral and ethical values that the person maintains. These can be religious, spiritual, or simply personal statements of value. Examples: Integrity, independence, compassion, respect, family, honor, God, patriotism, etc.

Physical traits – The physical characteristics that describe the person. Examples: Height, weight, appearance, attractiveness, fitness, athletic ability, etc.

Personality traits – The personality characteristics that describe the person. Examples: Likes change, Resists change, Likes company, Likes solitude, etc.

Goals – Everything the person wants to accomplish, not just career or family, but many values and interests also have corresponding goals. Examples: graduate with degree, get an A in chemistry, not compromise integrity, and not disrespect family, etc.

Experiences – The complete set of interactions the person has experienced. These include interactions with other people, with groups, with the natural world, and with himself. Examples: Successes, failures, education, and interactions, etc.

Interests and Hobbies – The set of subjects and activities that interests a person. Examples: Sailing, navigating, seashells, cooking, typing, and reading how-to books, etc.

Talents and Abilities – The innate talents and abilities that a person has. Talent is what we are born with. Examples: Musical talent, creativity, athletic ability, math ability, social talent, etc.

Skills – The actual levels of skill acquired in pursuing interests and developing talents. Skills are learned. Examples: Speaking skills, writing skills, etc.

Career – The career field the person has chosen to pursue, combined with current employment.

Family – The set of family relations that the person has. Mother, father, siblings, aunts, uncles, cousins, etc.

Aesthetic Preferences – Taste, sound, musical and artistic preferences, etc.

Friends and Associates – The people who are associated with a person. Her neighbors, friends, coworkers, etc.

Identity elements as quality plus quantity

Each identity element can be described as a qualitative description of the identity element, plus a quantitative measure of its strength.

For example, you may be especially gifted in math, and only moderately talented at writing. So you might consider yourself as "talented at math, 90 on a scale of 100" and "talented at writing 50 on a scale of 100."

The quantitative measure is only intended to show relative strength between identity elements within a single person. The measure is not intended to be comparative between different people. Therefore the measure only tells us that you "consider yourself better at math than at writing." Nothing more should be derived from the relative strengths of your identity elements.

The quantitative scale is purely arbitrary. You can think in a scale of 10, 100, 1000, 33, or whatever other number you prefer, so long as you are consistent.

Identity elements, as truths, are neutral

Truth is neither good nor bad, it simply is. Truth only becomes good or bad when viewed from a particular perspective.

Therefore, identity, as the collection of truths that define a person, is neither good nor bad. A person becomes good or bad only when viewed from a particular perspective.

A brief example: A wolf eats a farmer's sheep. The fact is neither good nor bad, only becoming good or bad when experienced from a particular perspective. From the point of view of a farmer, wanting to feed his family, it could be bad. But that is not the whole truth; rather it's only one perspective on the truth. From the point of view of the wolf, wanting to feed her pups, it is good. From the point of view of the sheep, preferring to remain alive, it is bad. Perhaps, from the point of view of a wild sheep, giving her life so her young can survive, it could be good. From the point of view of nature in general, it is neither good, nor bad, though at times of sheep overpopulation it might be good, and at times of wolf overpopulation it might be bad. Rather than fixate on any single one of these perspectives, the goal should be to see and accept as many of them as possible in order to have the most complete, most inclusive, most universal view of the truth.

One of the critical aspects of the model is that individuals must learn to accept the neutrality of truth, and they must learn that although the filter of their own perspective colors what they experience as positive or negative, the truth itself remains neutral. Truth has no agenda.

Positive and negative manifestation of identity

Every identity element can be expressed, manifested, and perceived with both positive or negative language and behavior.

As self-truths, however, identity elements are neutral. Even dishonesty can have both positive and negative expressions, as dishonest people can often be described as being diplomatic and good with people.

Negative	Neutral	Positive

TT Chart

TT charts demonstrate positive, neutral and negative language.

TT charts are useful for exploring the positive and negative aspects of all truths (not just personal identity elements) and finding neutral language to represent them.

By representing identity elements in a TT chart and finding positive, negative, and neutral language to express a single identity element, the neutral nature of self-truths can be explored.

Identity evolves through acceptance of truth

New identity is reinforced by retraining patterns

Building on the conscious vs. subconscious aspect of the model (presented later), we can conclude that the only way to truly grow is by accepting a new vision of self, then retraining the subconscious to teach it that we are no longer hurt by the old truth.

Every emotional crisis is really an identity crisis

Building on the conscious vs. subconscious aspect of the model, and the flow of acceptance and denial as a patterned process implemented in the subconscious, we can conclude that every emotional crisis, or neurosis, is really an identity crisis, caused by the conscious perceived identity diverging from the subconscious true identity.

The difference between perceived and actual identities within a person causes stress, denial, and irrational, neurotic behaviors.

The difference between perceived and actual identities between people causes stress, disagreement, and conflict.

The magnitude of difference between identity elements can be modeled with vectors.

Fundamental Model - Self-Esteem, Pleasure, Pain, Gain, Loss

Emotional pleasure is felt from positive experiences

Positive experiences are those that reinforce positively perceived identity elements or contradict negatively perceived ones.

If we aesthetically like blue skies, then blue skies will give us pleasure. If we believe we are good employees, then being recognized for a good job brings us pleasure.

Emotional pain is felt from negative experiences

Negative experiences are those that reinforce negatively perceived identity elements or contradict positively perceived ones.

If we believe that we are stubborn, then experiences that illustrate our stubborn nature are emotionally painful.

Experience is not always purely positive or negative

Sometimes an experience will trigger some positive and some negative perceptions of identity elements, causing conflicting emotions.

To understand conflicting emotions, decompose the experience to determine the affected identity elements, and then examine each separately.

If the affected identity elements do not adequately explain the emotional reaction to the experience, then keep looking for additional identity elements which were affected.

Self-esteem is gained by accepting positive inputs

This increase in self-esteem is the basis for the pleasure of the experience.

For example, if we as good employees are recognized as such, our self-esteem is increased.

Self-esteem is lost by accepting negative inputs

This decrease in self-esteem is the basis for the pain of the experience.

For example, if we as lazy slackers are recognized as such, our self-esteem is decreased.

Strong self-esteem is the acceptance of both true and perceived identity

Strong self-esteem (true self-esteem) comes from the acceptance of a positive view of both true and perceived identity

The only way to build true self-esteem is to discover and accept self truths.

In order to build self-esteem, the conscious has to believe that something is good, and the subconscious has to allow the conscious to experience it.

Weak self-esteem is the acceptance only of perceived identity

Weak self-esteem (false self-esteem) comes from the acceptance of a positive view of only perceived identity.

Declaring the truth into the conscious is a weak, temporary solution, one that ultimately will fail when contradictory evidence appears.

The use of behaviorist methods can accelerate the training of subconscious patterns, but if the patterns are not based on accepted self truths, then they will be weak and ultimately will fail.

Accept the negative aspects of positive attributes without pain.

Using TT charts to explore the positive and negative aspects of identity helps to unload negative language and experiences so we can learn to accept them.

Exploring the positive and negative aspects of the truth helpts o defuse irrational behaviors and distorted peceptions created by the subconscious to protect the identity from being hurt by negatively perceived aspects of the truth.

If we can accept negative inputs without pain, having resolved the dichotomy between positive and negative manifestations of single traits, then we can defuse the irrational behaviors developed to protect ourselves from accepting the negative aspects.

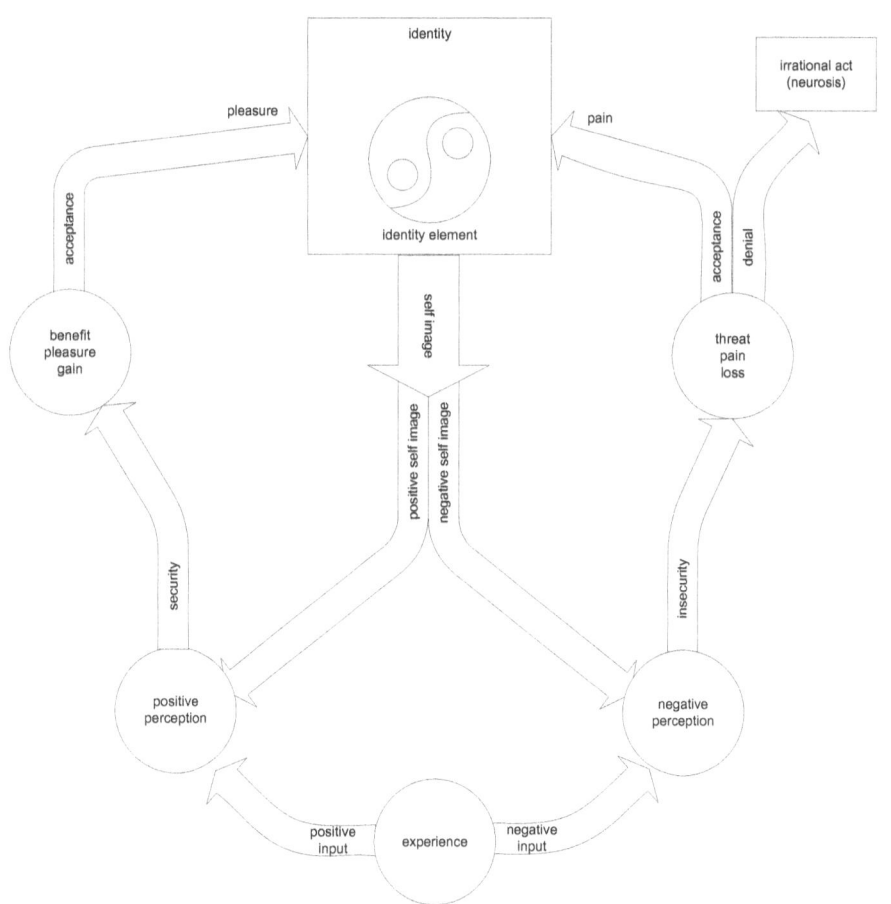

Identity, Acceptance and Denial

Experiences are interpreted as emotional pleasure or pain.

Gain of self-esteem is interpreted as emotional pleasure.

Loss of self-esteem is interpreted as emotional pain.

Our recognition pattern engine interprets a vast set of perceptions from every given experience. These perceptions are positive and negative depending on their affect on self-esteem.

The positive perceptions are allowed to pass on to the conscious, while the negative ones are distorted to make them positive (or at least tolerable).

Inputs which build self-esteem are accepted by the subconscious

When an experience triggers a perception that will produce pleasure by the conscious, it can be accepted with a high degree of accuracy.

Inputs which destroy self-esteem are denied by the subconscious

Inputs which are denied by the subconscious cause irrational action.

When an experience triggers a perception that will produce pain by the conscious, it is distorted so it won't cause pain.

The subconscious learns which interpretations of inputs produce pain and which produce pleasure purely by trial and error.

Acceptance of truth from accurate perception enables rational action

Action taken based on accurate perceptions is rational.

Distorted perception creates denial

Distortion occurs in perception, presentation, and in subconscious manipulation of information passed to the conscious.

Denial can take any number of forms: denial, blame, anger, projection, guilt, etc. But all manifestations of irrational behavior stem from denial at the root level.

Acceptance of falsity from distorted perception causes irrational action

Action taken based on distorted perception will result in irrational, neurotic action.

All neurotic actions are manifestations of denial.

Examples include guilt, blame, anger, excuses, projection, etc.

Repeated denial reinforces distorted patterns of perception and presentation

The more an irrational action is reinforced, the stronger and more intractable the distorted patterns become.

Irrational action can be directed inward or outward

Internal irrational actions can be in the form of guilt, shame, excuses, rationalization, blame, pity, etc.

External actions can be anger, blame, projection, denial, excuses, blame, outrage, offense, etc.

Emotional stress is caused by denial

Emotional stress results from the difference between perceived and true identity, which causes denial, insecurity, and anxiety.

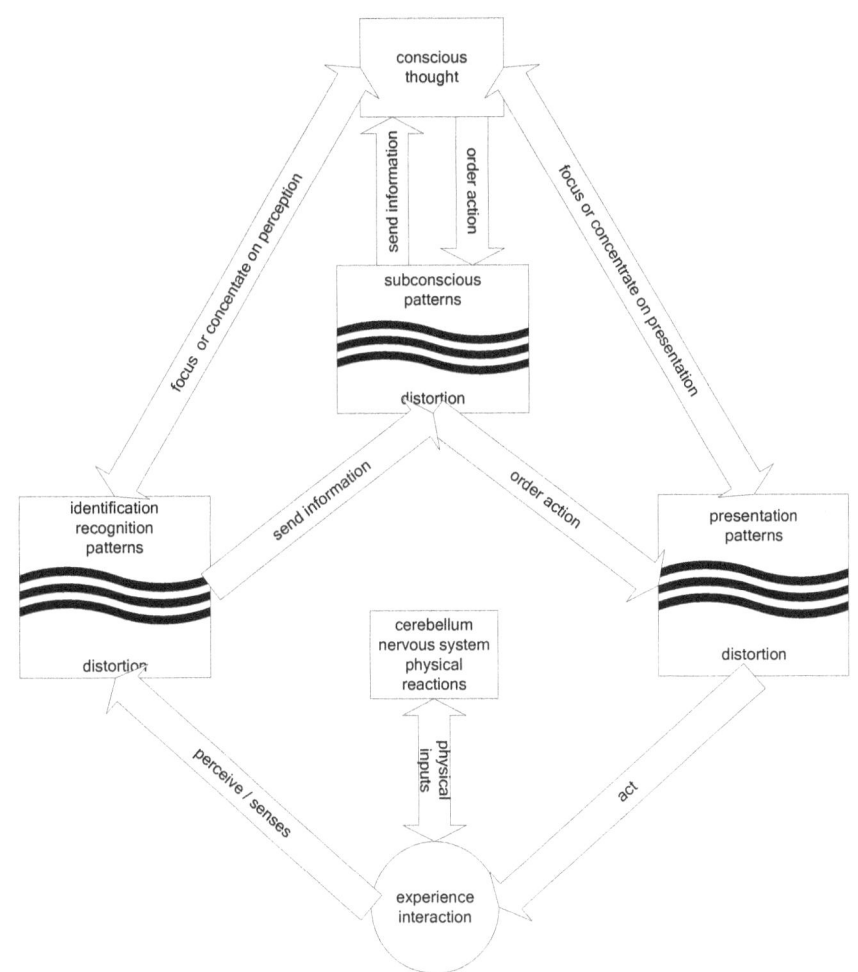

Cognitive Layers and Patterns

Conscious is always rational

Given accurate information, the conscious mind will always act rationally. However, the conscious is not given 100% accurate information, 100% of the time.

There are many opportunities for the introduction of distortion.

Conscious identity is perceived identity

It is our conscious self image, beliefs, values, etc.

Subconscious identity is true identity

It is learned entirely by experience and consists entirely of patterns that are applied to our perception and presentation to interpret experience and implement action.

The subconscious works by applying old interpretations to new data.

Conscious is not constrained by past patterns

The conscious is, however, strongly influenced by past patterns. This is because past patterns are applied to our perception by the subconscious, forcing conscious effort to perceive accurately in order to explore new thoughts.

The conscious is capable of reaching new conclusions with old information.

Conscious works only with perceived truths

Because the subconscious distorts perception based on perspective (true identity) using learned patterns, the conscious does not always have accurate information to work with. Garbage in leads to garbage out.

Only through conscious effort can accuracy of perception improve.

Conscious can focus on perception or presentation

The conscious has the ability to focus on the senses for perception and presentation.

Such attention requires explicit mental effort, and can only be applied to a limited scope and for a limited time. You can't pay attention to everything at the same time. And you can't pay attention to the same thing all the time.

However, focused attention is the only way to improve the accuracy of perception and presentation (through repetition).

Subconscious is entirely pattern based

The subconscious learns by example. It takes an input, and delivers or acts on it. If the action caused pleasure, then the pattern is encouraged, and next time the same pattern will be applied automatically. If the action caused pain, then the pattern is reinforced discouraged, and next time, a different interpretation may be tried.

Patterns are behaviorist in nature. They can be trained.

Subconscious learns to accept pleasure and deny pain

The subconscious is a neural network with only one feedback mechanism: emotional pleasure or pain.

The subconscious learns solely by experience using trial and error.

Subconscious acts independently and automatically

The vast majority of what we sense and what we do is managed subconsciously. Even high level tasks such as driving are often managed largely by the subconscious.

The vast majority of the time, the subconscious acts automatically without consulting the conscious. Unfortunately, it seldom informs it either.

The conscious is not usually explicitly aware of what the subconscious is doing, although traces do remain in short term memory, and can be examined with effort. The ability to review short term memory is key to evaluating the accuracy of patterns, and is an acquired skill that can be improved with practice.

Example: While walking, you approach a hole in the sidewalk. You automatically step over it without even noticing it is there. If you explicitly think about it moments later, you will remember stepping over the hole, while a short time later, you will probably not remember it at all.

Now consider that our emotional interactions are also highly patterned.

Subconscious knows true identity

The true identity exists only as the patterns learned by the subconscious. The subconscious learns to process information automatically and take automatic action the majority of the time. It learns to process the information using patterns. The patterns are developed as result of observing whether its actions cause pleasure or pain. Pleasure reinforces the pattern as valid, while pain contradicts the pattern as invalid.

Subconscious may rationally correspond with conscious

Subconscious may correspond with conscious and produce accurate perception and rational actions
When the conscious and the subconscious correspond, then the subconscious works rationally, and productively, producing rational actions, and delivering accurate information to the conscious.

Subconscious may irrationally vary from conscious

Subconscious may vary from conscious and distort perception and produce irrational actions.

When the subconscious and the conscious differ, the subconscious works irrationally, and counter productively, producing irrational actions and delivering inaccurate information to the conscious.

This discrepancy is the source of neurotic response.

Discrepancy between conscious and subconscious is the neurotic degree

For a given identity element, the difference between the actions taken by the conscious and the subconscious is the degree of neurosis.

Large discrepancies result in serious neuroses. Small discrepancies produce only minor neuroses.

Each individual has his or her own set of identity discrepancies, which result in his or her own set of neuroses. This, like all facts, is both bad and good. It's bad in that each of us has irrational elements of ourselves. It's good in that each of us has the ability to grow past them.

Can not query subconscious directly

The subconscious, existing only as a collection of patterns, can't be queried directly. It can only be inferred as the producer of the outputs (perceptions and actions) it produces for a given set of inputs.

Since it is seldom possible to test a single input at a time, and since the subconscious distorts what we see and do, deriving the subconscious identity is not trivial.

Subconscious can lie to conscious, but conscious can't lie to subconscious

The subconscious can distort perception and presentation, effectively lying to the conscious. This is done because it has learned through trial and error which particular interpretation the conscious prefers. This is often done to protect the conscious from the truth.

However, the conscious can't fool the subconscious. The subconscious always knows how its been taught to interpret the senses.

Change true identity by changing subconscious patterns

We can, having accepted a new identity consciously, wait for our subconscious to learn it gradually, over time.

We can also apply behaviorist therapies to accelerate the retraining of our patterns.

Achieve long lasting changes in patterns through training

By accepting new truths, and training evolved conscious identity into subconscious patterns, you achieve long lasting changes. Retraining old patterns is the only route to permanent growth.

Behaviorist methods can be applied to accelerate the retraining of old patterns.

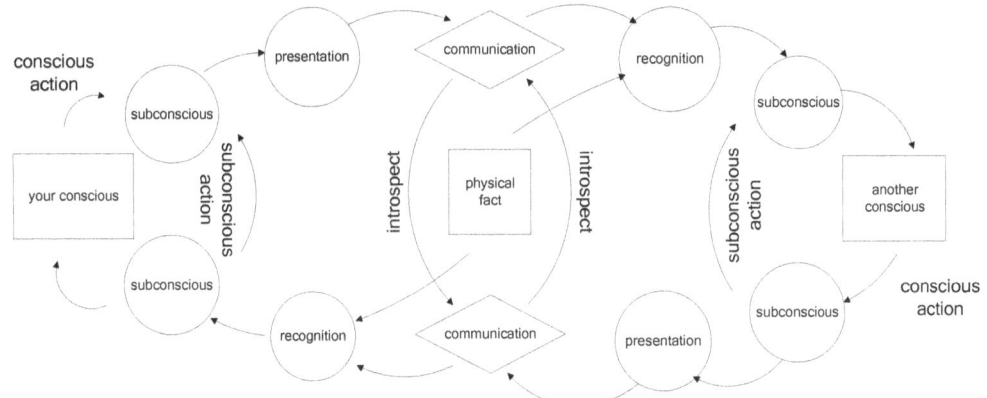

Perception and Presentation

Subconscious distorts perception and presentation

Based on anticipated pleasure or pain resulting from differences in true and perceived identity and established patterns, the subconscious will distort perception and presentation without your knowledge

Accurate perception and presentation is an acquired skill

Your perception is a set of skills. You can develop your perception by training your patterns to transmit truth accurately. In order for this to be possible, the truth must not cause emotional pain, else the subconscious will learn to only deliver it well distorted.

Your presentation is also a set of skills. You can improve the accuracy of your presentation through continuous practice.

By paying careful attention, practicing, and training while observing the level of accuracy in your performance, you can train the subconscious perception and presentation.

We can grow randomly and haphazardly, or we can adopt a philosophy of growth and learning where we actively monitor our emotional state and continously improve the accuracy of our patterns both for perception and for presentation.

Derived Facts - Emotions, Thoughts

Emotions are subconscious response to anticipated impact on self-esteem

When your subconscious predicts the effect an experience will have on your self-esteem, it manifests behavior, feelings, and actions which are the basis for emotions.

Thoughts are conscious response to anticipated impact on self-esteem

When your conscious predicts the effect an experience will have on your self-esteem, it manifests behavior, feelings, and actions with are the basis for thought.

Derived Facts - Love

Love is acceptance of true identity

Love is built on the acceptance of a person.

Love of another is acceptance of another's identity

Love of another is the acceptance of that person for who they are believed to be, to the extent that it is known.

Love by others is the acceptance of ones identity by others

Love by another is their acceptance of a person for who they are believed to be, to the extent that it is known.

Self-Esteem is the acceptance of oneself

Love of self, or self-esteem, is the acceptance of one's perceived identity, as reconciled with one's true identity.

Strong self-esteem must be built on an acceptance of the truth into one's identity.

Weak self-esteem can be temporarily built by declaring false truths into the one's identity, however reality will eventually break through. This is done through various forms of denial, and causes insecurities and irrational or reactionary behavior.

Derived Facts - Rejection, Failure, Mistakes

Rejection is an experience contradicting identity

Rejection by another results in a loss of self-esteem as either a contradictory input of a positively perceived identity element (ie no you are not interesting) or as a confirming input of a negatively perceived identity element (ie you are boring).

Failure is an experience contradicting identity

Failure results in a loss of self-esteem as it is the acceptance of a contradictory input of a positively perceived identity element (you did not achieve your goal).

Mistakes are experiences contradicting identity

Mistakes represent failures on a smaller scale. Each mistake is a failure of the goals of achieving perfection and being right.

Derived Facts - Happiness, Sadness

Happiness is pleasure combined with satisfaction

True happiness is both the pleasure of the moment combined with an overall sense of satisfaction.

Pleasure is the gain of self-esteem

Pleasure results from the reinforcement of positive identity elements or the contradiction of negative ones.

The pleasure of the moment is the acceptance of any experience which produces self-esteem.

Pain is the loss of self-esteem

Pain results from the reinforcement of negative identity elements or the contradiction of positive ones.

The pain of a moment is the acceptance of any experience which reduces self-esteem.

Sadness is the acceptance of pain

Accepting pain, allowing the psyche to accept and realize a loss, is manifested as sadness and as a loss of emotional energy.

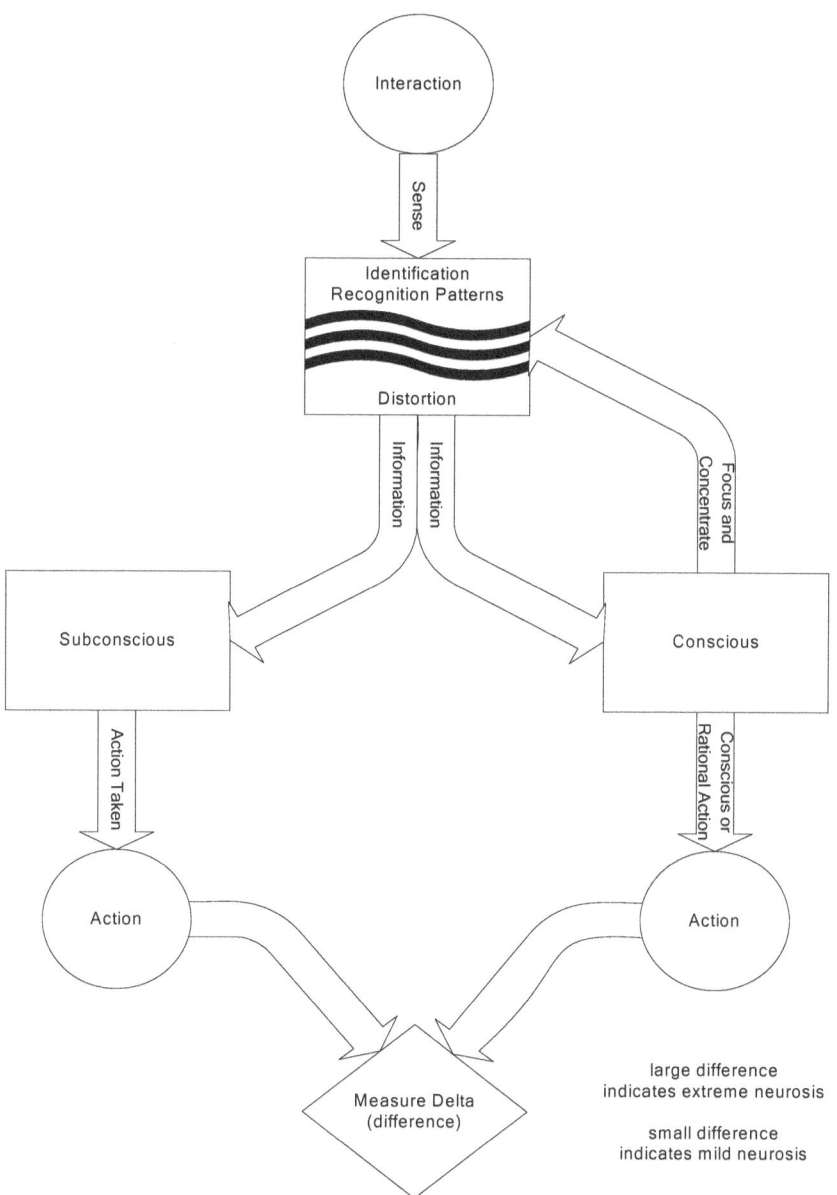

Measuring Neurotic Degree

Conscious identity is perceived identity

The conscious identity is not the true identity. It is the perception of identity, or the desired identity.

If you think of yourself as a loving person, then that is part of your conscious identity. If you think you like classical music, then that is part of your conscious identity. And so on for each element of the consciously accepted truths that define you.

Subconscious identity is true identity

To the degree that the subconscious processes the vast majority of our experiences, and manages the vast majority of our actions, it also defines the vast majority of our true identity.

If you think of yourself as a loving person, but in reality you never actually make time for anyone but yourself, then your subconscious may be implementing a pattern of self indulgence and realizing a true identity of being less loving than you believe yourself to be.

Because this is happening in the subconscious, the subconscious will also conveniently protect you from having to realize that it is happening at all.

Only by paying attention to actions, and studying your own behavior, reactions, and emotions will you be able to discover the ways in which your true identity is not aligned with your perceived identity.

Discover true identity through examination of subconscious behavior

Examination of subconscious can only be done by examining subconscious actions (outputs/presentation) based on certain experiences (inputs/perception) and comparing them to hypothetical conscious actions based on the same inputs.

Learn to find and explore our patterns

Only by exploring our subconscious patterns, our personal perspective and our presentation filters, can we discover our true subconscious identity.

Learn to accept true identity

Only by accepting the truth of our identity, can we gain the power to change it.

Learn to evolve true identity

Through retraining of patterns, eliminate obsolete identity elements and forge new ones based on new truths.

Discover new truths through philosophy and self discovery.

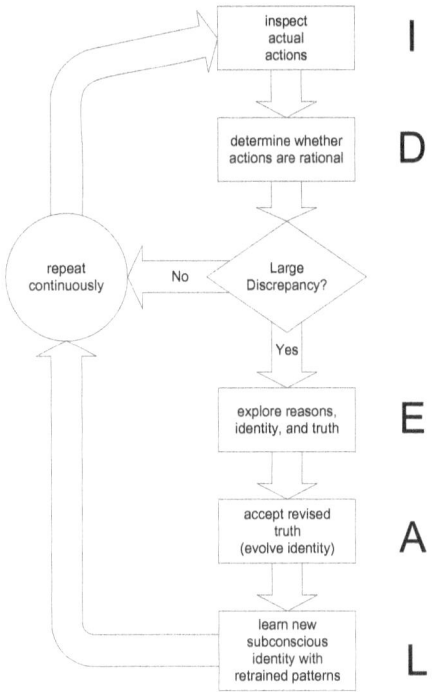

Process of Personal Growth

Inspect actual behavior, actions, and feelings

Observe your behavior for signs of denial.

Identify hot buttons and insecurities in order to observe when they are triggered.

Determine rationality of action

Identify neurotic degree by comparing actual behavior, action or feeling with rational ones.

Explore reasons, identity, and truth

Explore the cause of the insecurity or neurosis that caused the irrational behavior, action or feeling.

Explore expanded or alternate truths that would relieve the insecurity.

Accept the evolved truths into perceived identity

Must be able to accept and experience both pleasure (gain, success) and pain (loss, rejection, failure, mistakes) honestly before you can retrain the subconscious effectively.

Learn the new identity in the subconscious

Retrain the subconscious in order to affect long lasting personal growth.

Subconscious can be left to retrain gradually with new perceived identity.

Subconscious retraining can be accelerated using behaviorist methods.

Defuse hot buttons, insecurities, and irrational reactions by discovering truth, evolving identity elements, accepting new identity elements, and by retraining patterns using behaviorist methods.

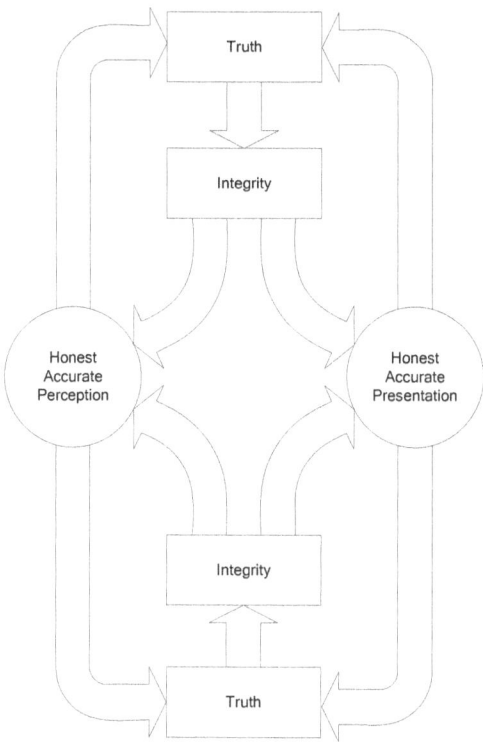

Truth, Integrity and Honesty

Truth is obtained through honest perceptions of experiences

Truth depends on the honesty and accuracy of our filters.

To perceive the truth, we must build accurate perceptions. This requires effort and attentiveness.

To present the truth, we must build accurate presentation habits.

Integrity is the commitment to honesty.

Integrity is the personal commitment to and direction of effort towards building and maintaining honest, accurate perception and presentation filters.

Left to it's own devices, the subconscious will only learn to not deliver painful information. But it has no way of knowing if it is accurate in its judgements.

Explore truth through philosophy

Learn philosophical methods to apply in the search for truth and reason.

Socratic Questioning

Accept no truth as absolute

Few truths are completely knowable (Kant)

Synthesis: Blend contradictory truths to find inclusive truth

Study and explore other philosophical methods and schools of thought.

Nothing is really true unless it is universal

Truth must be true from all perspectives to be universal.

Any truth that is contradicted by a conflicting point of view has to be amended to accomodate the existence of the conflicting point of view before it can be valid.

Evolving a partial truth by restricting its context is one way to make it more universal.

Seek original sources

While summarized and commonly presented knowledge may be useful as a starting point, when exploring truth, always seek the sources closest to the origin of the truth. Each person that information passes through between the source and the recipient is an opportunity for distortion.

Relying on second or third hand information can be dangerous, as you are vulnerable due to your thin, poorly developed first hand knowledge.

Be aware of the difference between what you know and what you have heard.

Be aware of the difference between what you think and what you have proven.

Be aware of the scope and comprehensiveness of your proofs.

Seek and explore disparities

Learn to identify disparities, exceptions, contradictions, and omissions as clues that beg further investigation.

Seek and explore alternate perspectives

When faced with opposing views, try to state the rationale behind the opposing position. This is an aid to understanding both for you, and also for the other person, as it helps them to observe the accuracy of their presentation.

Seek and explore supporting and contradictory facts

Seek all the facts, not just those that support your position. If your truth is weak and insupportable, you are better off challenging it yourself.

Examine inward and outward actions

Each action, behavior, and feeling should be questioned for rationality.

Each irrational action, feeling, and behavior identified indicates an underlying insecurity and identity discrepancy, which is an opportunity for personal growth.

Examine stressors

Every source of emotional stress indicates an underlying insecurity and identity discrepancy, which presents an opportunity for personal growth.

You can only see yourself through interactions

Learn to observe yourself through the eyes of others.

Learn to hear what people really say, rather than what you want to hear.

Learn to hear what people really mean, rather than what they say.

Practice and train your presentation

Our communication skills, both verbal and non-verbal are emtotional patterns implemented in the subconscious.

Your presentation is a set of subconscious patterns which you learned using trial and error from feedback. It can be retrained with sufficient effort. An athlete practicing swinging a bat is a physical example of training presentation, however emotional patterns work in a similar manner. A child learning to walk is another example. On an emotional level, joining Toastmasters could be a behaviorist method for training presentation (in the context of speaking in front of groups).

Practice, practice, practice. Then practice some more.

Practice and train your perception

You must also train your perception. By identifying inaccuracies in your perception, and concentrating on seeing past them, you can retrain your perception to be more accurate.

By learning to identify your hot buttons (by identifying irrational reactions), and training your presentation to take a pause before responding, you can reinsert a critical element of control in your life.

It all starts with perceiving the behavior of the subconscious, and training the subconscious to perceive accurately.

Learn to recognize filtering through subtle traces in short term memory

Just as you can remember, briefly, that you stepped over a toy in the living room, you can also briefly remember your subconscious emotional actions as well.

As a tool for discovering your true identity, practice observing your subconscious in operation. You can't change who you really are tomorrow until you first identify and accept who you really are today.

Learn to defuse loaded language

Loaded language is the presentation of truth in a positive or negative wrapper. It can be done subconsciously or consciously, and it affects both perception and presentation.

When presenting, observe and correct your tendencies with respect to subconscious use of loaded language. Strive to use loaded language only with awareness and only to achieve the specific effect you intend.

When perceiving, learn to identify loaded language and take the spin off the ball. Learn to desensitize your hot buttons and to see truth left behind after removing the intent of the other party, rather than any injury or harm to yourself.

While someone else can hurt your body, only you can let your feelings be hurt. Offense can only be taken, not given.

Learn to target presentation past perspective

When presenting, learn to identify the recipient's perspective and to target your presentation specifically to get past it.

For example, a coworker doesn't care about the boss' point of view so citing the boss' opinion will likely not be persuasive. Likewise, a boss doesn't generally care about a coworker's feelings, so complaining about someone's hurt feelings to the boss is not likely to accomplish productive results.

Instead, learn to repackage information to target the perspective of the recipient. By packaging a directive to point out benefits to a coworker, the coworker is more likely to accept and follow the directive. Likewise, by repackaging a coworker's complaint in terms of lost productivity and wasted money, a boss is more likely to consider a complaint she may otherwise have dismissed as sour grapes.

Learn to identify and control button pushing actions

Learn to identify behavior and presentation that pushes other people's buttons.

Retrain your presentation to not push those buttons subconsciously.

Remember what you used to do, as it can be effective to push buttons intentionally. But you should always strive to be 100% aware of everything you do, and the reasons why you do it.

www.ingramcontent.com/pod-product-compliance
Lightning Source LLC
Chambersburg PA
CBHW021301280526
45784CB00005B/2470